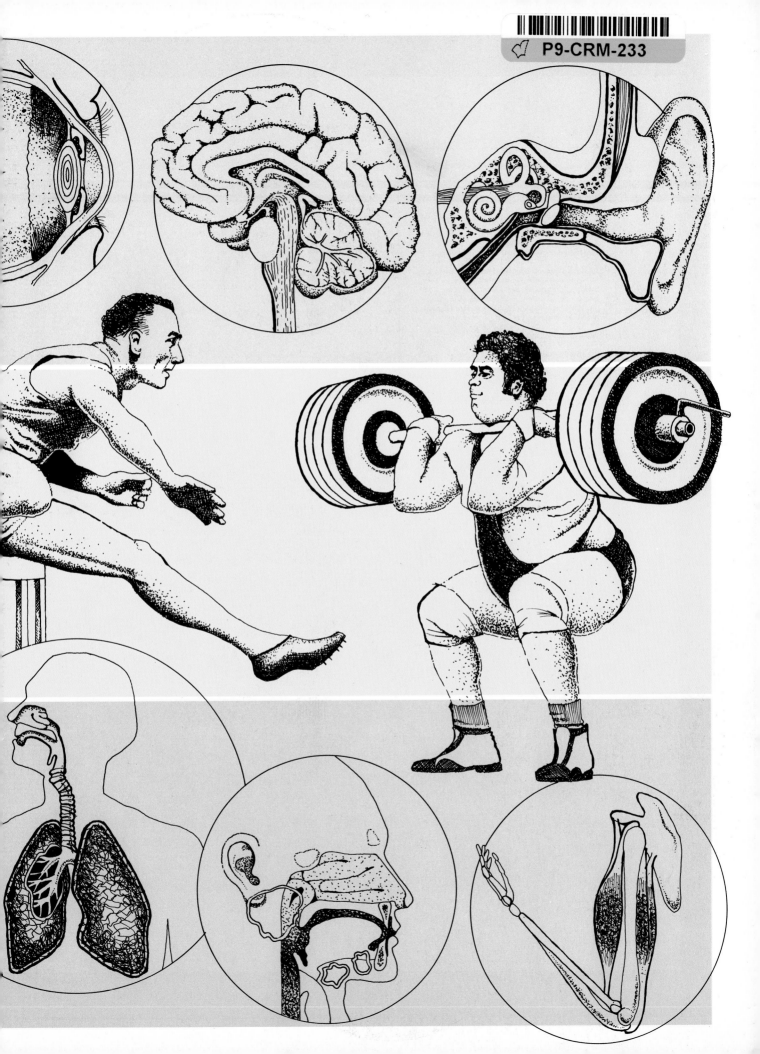

First published in 1989 by
Treasure Press
Michelin House
81 Fulham Road
London SW3 6RB

ISBN 1 85051 355 4

Produced by Mandarin Offset
Printed and bound in Hong Kong

First Questions
about the
HUMAN BODY

Keith Faulkner

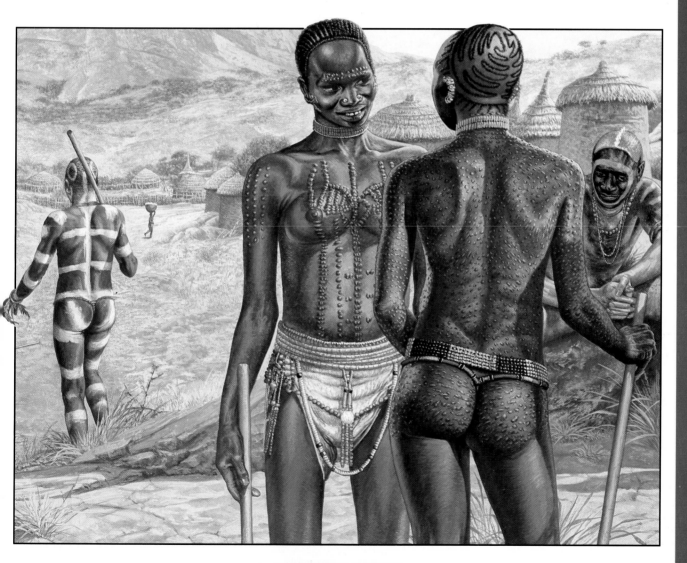

TREASURE PRESS

Conceived and Produced by:
Keith Faulkner Publishing Limited

Designed by Jonathan Lambert

Illustrated by:
Liz Graham-Yooll
Richard Hook
Clifford & Wendy Meadway
David Palmer
Bernard Robinson
Mike Roffe
John Sibbick

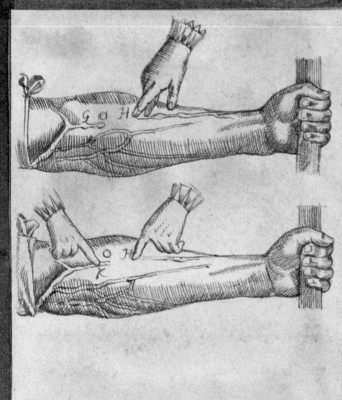

Contents

How did man evolve?	8
Sorcery or science?	10
Who made discoveries?	12
What's inside your body?	14
What are you made of?	16
How many bones?	18
What makes muscles?	20
Why is blood red?	22
Can we live without air?	24
What fuels your body?	26
What are chemical messengers?	28
Are you electric?	30
Is the brain a computer?	32
What has rods and cones?	34
Do ears help you balance?	36
What has 10,000 buds?	38
Why do we chew?	40
What gives 2 sq m of protection?	42
Did you start life as an egg?	44
Do you look like your parents?	46
Why are people different colours?	48
How does the body defend itself?	50
Can people be repaired?	52
Can Man compete with animals?	54
Do we have supernatural powers?	56
What can the body achieve?	58

How did man evolve?

There is no real beginning to the story of Man, except the beginnings of life on the Earth about 700 million years ago.

Man's earliest true ancestors, the primates, evolved about 60 million years ago; these were tiny squirrel-like creatures. The first man-like primates were like the chimpanzee and lived in a similar way. One million years ago *Homo erectus*, or Upright Man, looked almost like modern man.

Our knowledge of early Man is based on the discovery of fragments of bone and tools. As this detective work continues, new finds will increase our knowledge, but we may never really be sure of our earliest ancestors.

Early Man's hands played an important role in his development. The ability to grip objects between fingers and thumb gave him a great advantage for making and using tools.

Life in the Neolithic (New Stone) Age had become fairly organised. The men hunted in groups to capture large animals, using beautifully made weapons of flint.

The women stretched and prepared animal skins which were sewn together with bone needles. Fire was very important, for warmth and protection from dangerous animals. It was tended carefully and the glowing embers were carried from place to place in clay-lined baskets.

The Bigfoot or Sasquatch of North America has been seen by hundreds of people, and even filmed once. Yet no specimen has ever been captured. Could this be a relative of ours, surviving in the otherwise uninhabited forest?

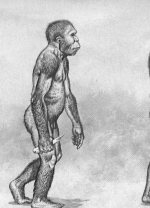

Australopithecus Homo Habilis Homo Erectus Neanderthal Homo Sapiens

We cannot really be certain about the evolution of Man. It is thought that *Homo erectus* and Neanderthal man may have lived at the same time and competed with each other for food.

Sorcery or science?

Since the earliest times, religion and medicine have been closely connected. Even today there are many primitive tribes who rely on the witch doctor to cure their ills. Many of the witch doctor's methods of treatment may seem very strange, but they are of interest to science as, in some cases, they do seem to work.

In the forests of South America, a tribe was discovered that seemed to have an effective way of treating wounds. A certain plant was first chewed and then placed in a gourd and hung from the branches of a special tree. After a few days a whitish mould appeared, and this was placed on the wound and bound with leaves. When the mould was analysed it was found to be penicillin.

The Chinese have used acupuncture for thousands of years. It is used for treating illnesses and even as an anaesthetic.

In the 15th century, Leonardo da Vinci was making careful scientific study of the human body. He dissected corpses and made detailed drawings and notes of his findings.

The Indians of North America had great faith in the medicine man's ability to cure illness and even insanity. Wearing a mask or animal head-dress, the medicine man used a mixture of magic charms and possibly herbal medicine to cure his patients or drive out evil spirits.

Some people believe that they have the power to heal, just by placing their hands on the affected part of someone's body. In many cases people can be healed if they really believe that this is possible.

Who made discoveries?

Before medieval times in Europe, illness and disease were thought to be caused by evil spirits. The fear of death and the dead prevented people from trying to discover the workings of the human body.

It was not until the 15th century that real progress was made in the field of anatomy. The Pope gave his permission for human dissection, and in England, Henry VIII allowed the bodies of four criminals per year to be used for anatomical studies. There was still a shortage of available bodies, and the illegal services of grave robbers were used by many early anatomists.

During the 16th century many European universities had a school of medicine. The students would learn about the anatomy of the human body by watching dissections in an anatomy theatre.

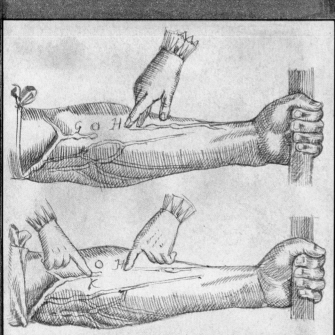

In 1628, William Harvey published an accurate plan of the body's blood circulation. As there were no microscopes, he could not have known about the minute capillaries that join the arteries and veins.

The first book on human anatomy was published in the 16th century by Vesalius. It contained many detailed drawings based on the dissected corpses of executed criminals.

Two 18th century grave robbers, Burke and Hare, are believed to have murdered 15 people in order to sell their bodies to anatomists.

What's inside your body?

The human body is an incredibly complex machine, controlled by a brain more powerful and versatile than the largest computer.

The body is composed of many parts – the skeleton providing a strong but flexible framework – and is moved by the contraction of muscles controlled by the brain. The heart, the lungs and the blood system provide oxygen and carry away waste material, whilst the teeth, salivary glands and digestive system break down food into the fuel that drives the body machine.

We are only just beginning to understand how our bodies work and, most importantly, how to keep them working efficiently. But although our knowledge is growing rapidly, the human body will always remain a marvellous and mysterious machine.

The simple action of hitting a tennis ball requires co-ordination of many parts of the body. The eyes and brain judge the distance and speed of the ball, whilst the inner ear and muscles balance the body.

— Oxygen

— Carbon

— Hydrogen
— Nitrogen
— Calcium
— Phosphorus
— Others

The human body is composed of many different chemical elements. It may be surprising to discover that 70% of your body is water.

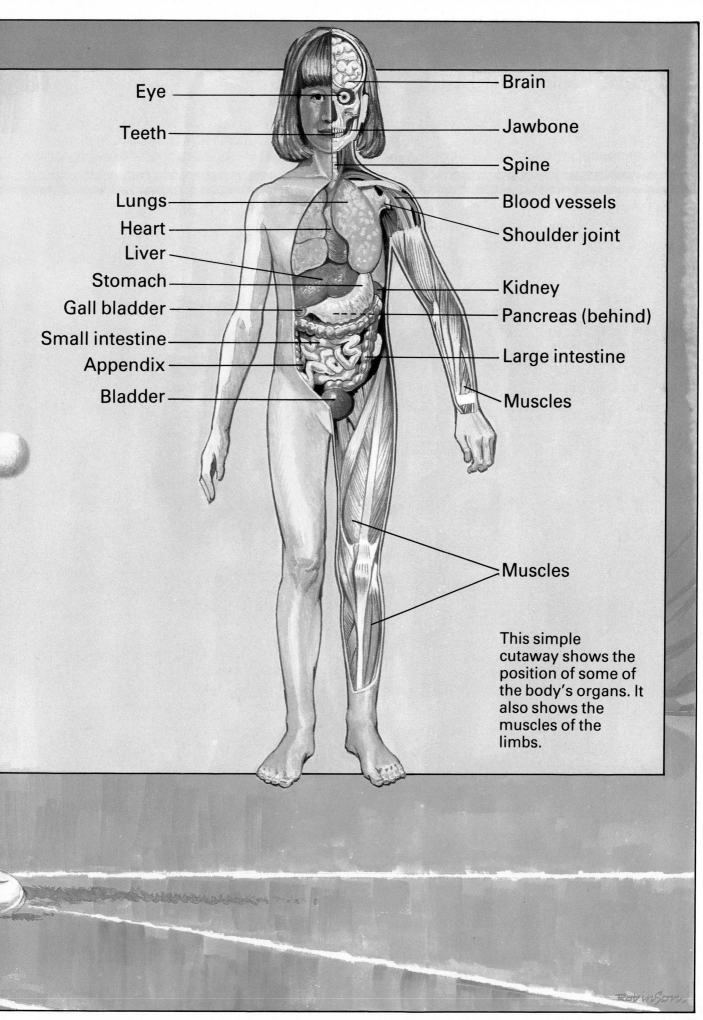

Eye — Brain

Teeth — Jawbone

— Spine

Lungs — Blood vessels

Heart — Shoulder joint

Liver

Stomach — Kidney

Gall bladder — Pancreas (behind)

Small intestine — Large intestine

Appendix —

Bladder — Muscles

Muscles

This simple
cutaway shows the
position of some of
the body's organs. It
also shows the
muscles of the
limbs.

What are you made of?

For thousands of years, Man's knowledge of the human body was limited to what could be seen with the naked eye. In the 17th century, Robert Hooke discovered that living matter was composed of many tiny parts, which he named cells.

Your body may contain about 50 billion cells. Some cells fight germs, some are grouped together to form bones or organs, and others are chemical factories.

Our knowledge of cells and how they work has increased with the development of the microscope. The electron microscope is powerful enough to show the structure inside each cell.

This illustration shows a typical human cell. Although cells may vary in size and shape, they all contain the same basic parts. These details are only visible by using an electron microscope to magnify the cell to 20,000 times its actual size.

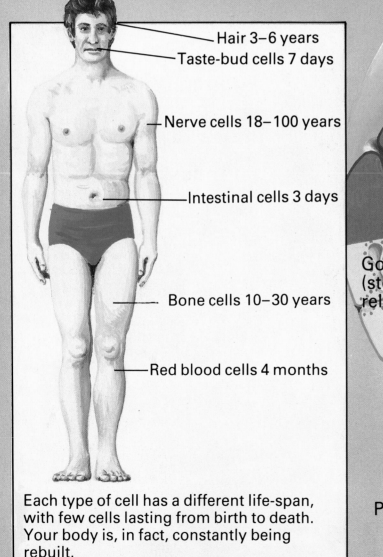

Hair 3–6 years
Taste-bud cells 7 days

Nerve cells 18–100 years

Intestinal cells 3 days

Bone cells 10–30 years

Red blood cells 4 months

Each type of cell has a different life-span, with few cells lasting from birth to death. Your body is, in fact, constantly being rebuilt.

Cell membrane (outer skin)

Golgi apparatus (store chemicals for release from cell)

Centrosome (controls cell division)

Protein factories

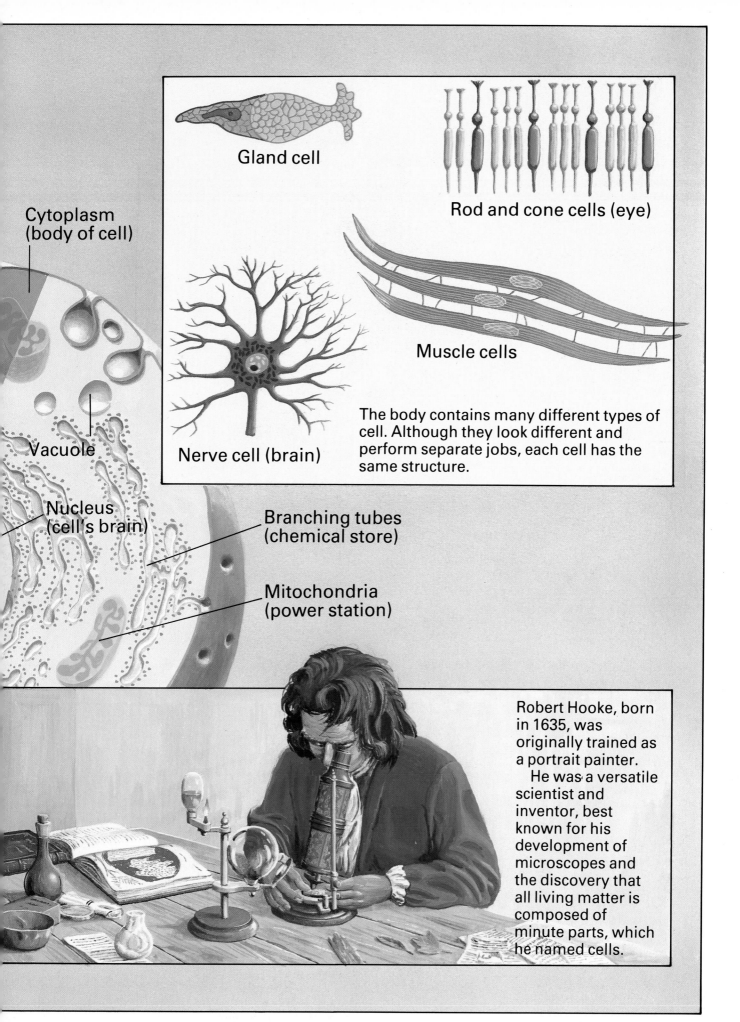

Gland cell

Rod and cone cells (eye)

Cytoplasm
(body of cell)

Vacuole

Nucleus
(cell's brain)

Branching tubes
(chemical store)

Mitochondria
(power station)

Nerve cell (brain)

Muscle cells

The body contains many different types of cell. Although they look different and perform separate jobs, each cell has the same structure.

Robert Hooke, born in 1635, was originally trained as a portrait painter.
    He was a versatile scientist and inventor, best known for his development of microscopes and the discovery that all living matter is composed of minute parts, which he named cells.

How many bones?

We may not think of ourselves as walking masterpieces of engineering, but the human skeleton is a mechanical miracle.

The 206 bones which form the skeleton provide us with a supporting framework that has a flexibility and range of movement far greater than any mechanical device. The skeleton is a very versatile structure, comprising bones, cartilage, tendons and ligaments. Each part of this structure is designed for a special job; the skull and the ribs protect the delicate brain and internal organs, and the spinal column supports the body, provides movement, and protects the spinal cord from damage.

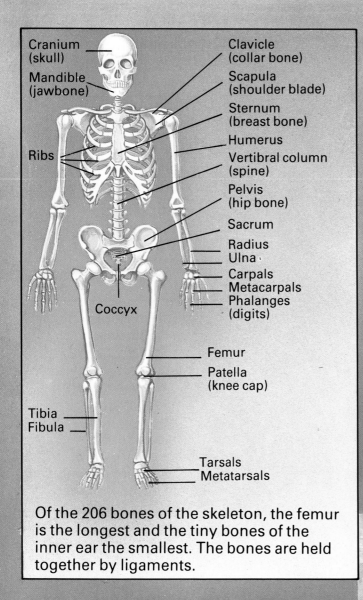

Cranium (skull)
Mandible (jawbone)
Ribs
Coccyx
Tibia
Fibula

Clavicle (collar bone)
Scapula (shoulder blade)
Sternum (breast bone)
Humerus
Vertibral column (spine)
Pelvis (hip bone)
Sacrum
Radius
Ulna
Carpals
Metacarpals
Phalanges (digits)
Femur
Patella (knee cap)
Tarsals
Metatarsals

Of the 206 bones of the skeleton, the femur is the longest and the tiny bones of the inner ear the smallest. The bones are held together by ligaments.

The flexibility of the human skeleton is demonstrated by the movement of a gymnast or dancer. It is only by constant practice that the ligaments are elastic enough to provide this degree of graceful movement.

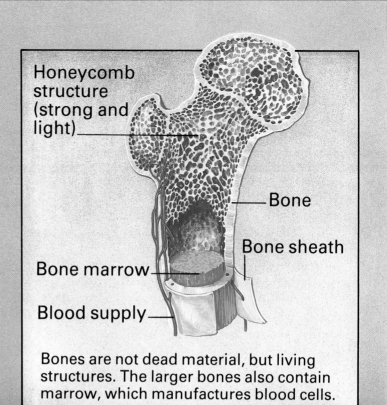

Honeycomb structure (strong and light)

Bone

Bone sheath

Bone marrow

Blood supply

Bones are not dead material, but living structures. The larger bones also contain marrow, which manufactures blood cells.

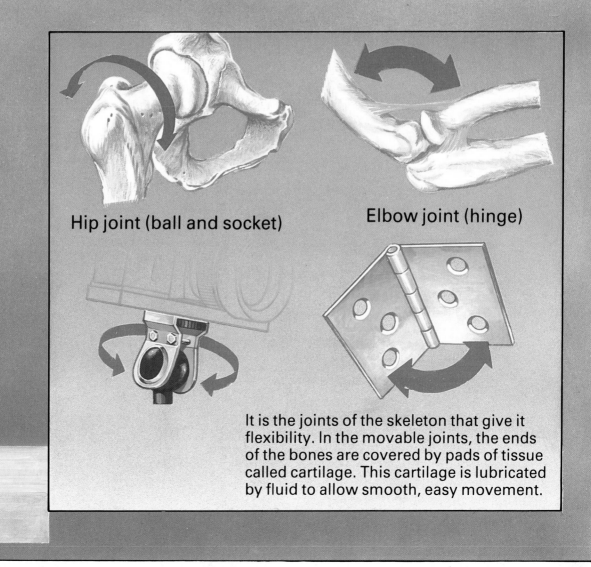

Hip joint (ball and socket)

Elbow joint (hinge)

It is the joints of the skeleton that give it flexibility. In the movable joints, the ends of the bones are covered by pads of tissue called cartilage. This cartilage is lubricated by fluid to allow smooth, easy movement.

What makes muscles?

Whether you are lifting heavy weights in the Olympics, or simply blinking your eye, you are using muscles. About 620 muscles control the movements of your body, and another 30 hidden muscles are working automatically. These automatic muscles control your digestion, heartbeat and many other internal functions which you don't need to think about.

There are three main types of muscle – skeletal (or striated) muscle for voluntary movement, smooth (or involuntary) muscle which lines the intestines, and heart (or cardiac) muscle. All muscles have one thing in common, the more they are used, the stronger they become.

The weight lifter below has developed strong muscles by regular training.

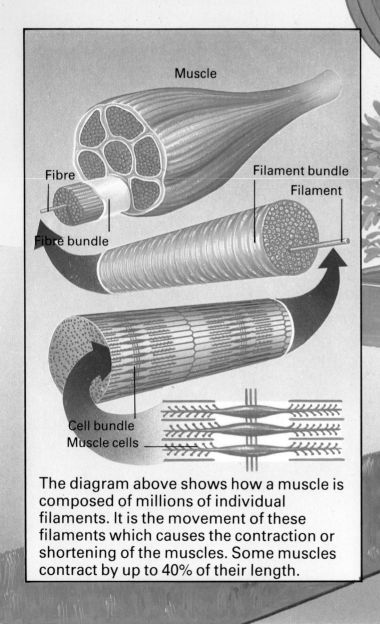

Muscle

Fibre

Fibre bundle

Filament bundle

Filament

Cell bundle

Muscle cells

The diagram above shows how a muscle is composed of millions of individual filaments. It is the movement of these filaments which causes the contraction or shortening of the muscles. Some muscles contract by up to 40% of their length.

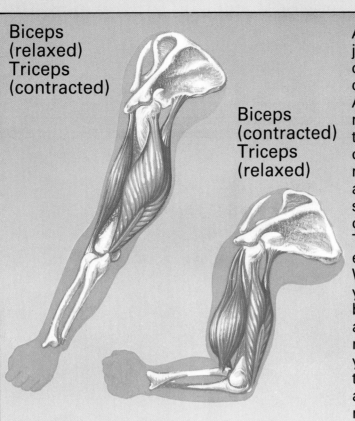

Biceps
(relaxed)
Triceps
(contracted)

Biceps
(contracted)
Triceps
(relaxed)

A movement, like jumping, may use over a hundred different muscles. Almost all movement requires the use of a group of muscles. Often muscles are arranged in pairs, so that they work in opposite directions. The simplest to explain is the arm. When you bend your arm, the biceps contracts and the triceps relaxes, but when you straighten it, the triceps contracts and the biceps relaxes.

Even when you are asleep, some of your muscles are still working. They are called involuntary muscles because they work automatically. Your heart, lungs and digestive system all have involuntary muscles.

Why is blood red?

Every day, the heart of a normal adult pumps about 9,000 litres of blood through almost 100,000 km of circulatory system. This incredible network of tubes reaches every cell of the body, carrying nutrients and oxygen and taking away carbon dioxide and waste products for disposal.

The average adult has about 4–5 litres of blood, which completes each circuit of the body in less than a minute. Blood contains many substances, but the most important are the red cells and the white cells. The red cells get their colour from a pigment called haemoglobin. Without this, blood would be a clear yellowish liquid. In every single drop of blood there are about 250 million red cells, which carry oxygen. The white cells are the body's police force, they attack and destroy germs and bacteria. Although they are larger and less numerous than the red cells, there are still about half a million in each drop of blood.

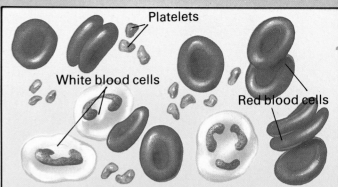

Blood is a liquid called plasma with red cells which carry oxygen, white cells that attack bacteria and platelets to induce the blood to clot.

The rapid heartbeat and gasping breaths at the end of a race are caused by the body trying to replace oxygen.

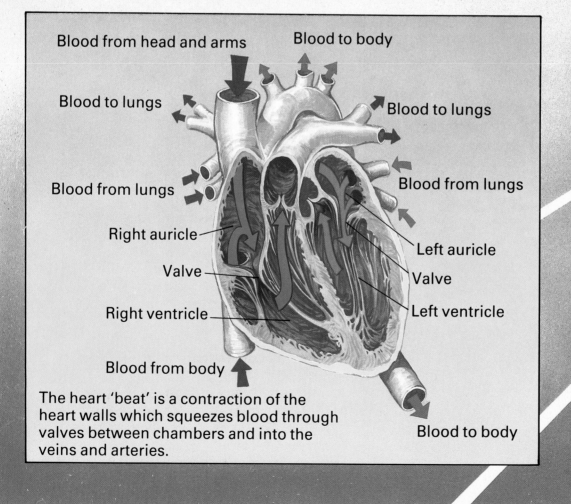

The heart 'beat' is a contraction of the heart walls which squeezes blood through valves between chambers and into the veins and arteries.

The diagram on the right shows the blood circulation – arteries are in red and veins in blue. The blood is pumped from your heart into the arteries; from there it travels to every part of your body in fine tubes called capillaries. When the oxygen is used, the blood collects carbon dioxide and waste material, and travels through the veins back to your heart. The heart then pumps it to the lungs to expel carbon dioxide and collect more oxygen.

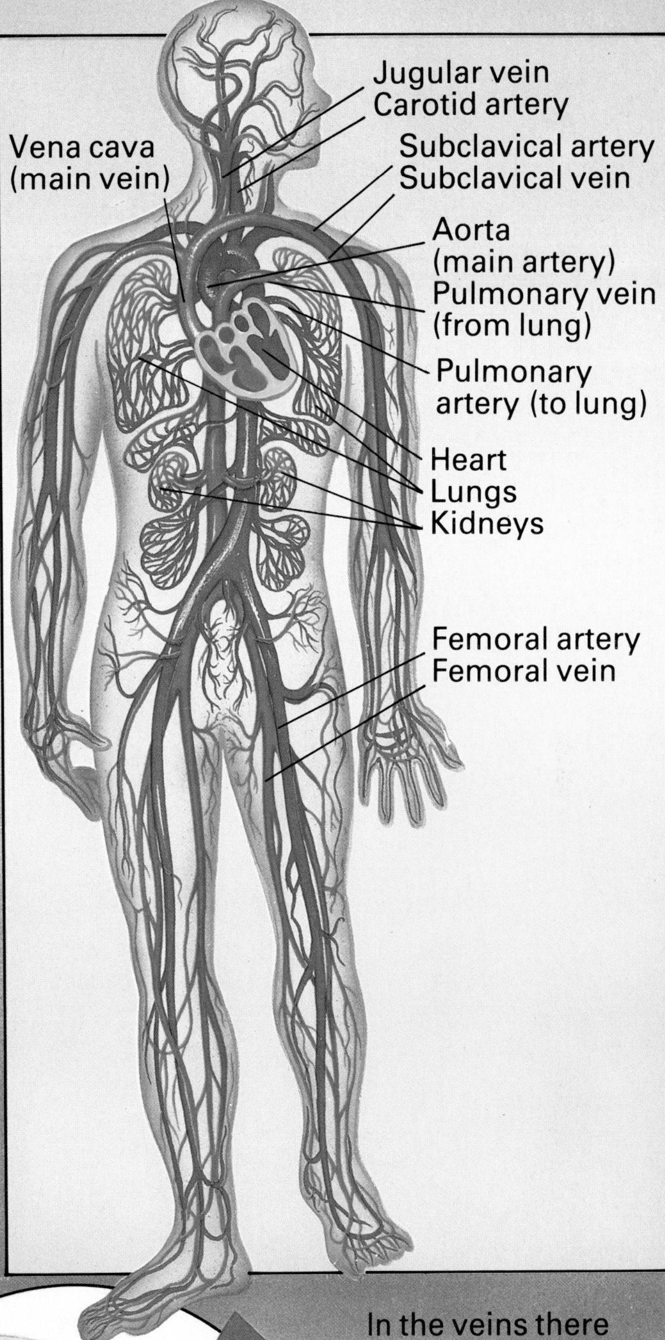

Vena cava (main vein)

Jugular vein
Carotid artery
Subclavical artery
Subclavical vein

Aorta (main artery)
Pulmonary vein (from lung)

Pulmonary artery (to lung)

Heart
Lungs
Kidneys

Femoral artery
Femoral vein

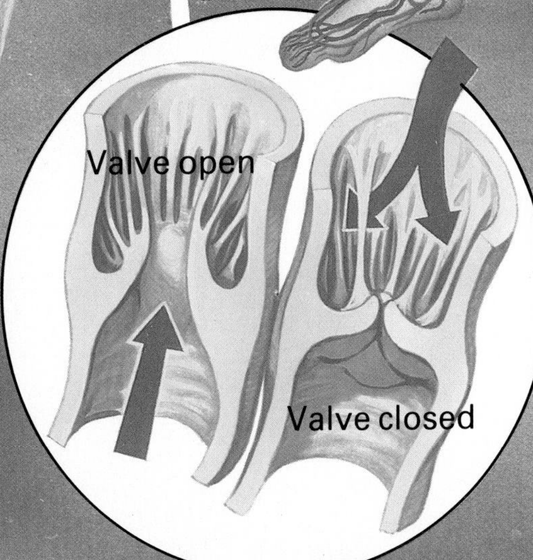

Valve open

Valve closed

In the veins there are special valves which only allow the blood to flow in one direction. Each beat of the heart pushes the blood along the veins and the valves prevent it flowing back between the 'beats'.

Can we live without air?

The human body can survive without food for several weeks. It can go without water for several days, but if it is deprived of oxygen for just a few minutes, it will die.

The body is like a very slow fire, it consumes food as fuel, and uses oxygen to 'burn' it. The result is energy, which is like the heat of the fire, and the waste product is carbon dioxide. The more energy the body needs, the more fuel it requires and more oxygen to burn it. This entire process is called respiration.

You can hold your breath for a minute or two, but very soon you begin to feel the need to breathe. The diver below overcomes this problem by carrying tanks of oxygen on his back.

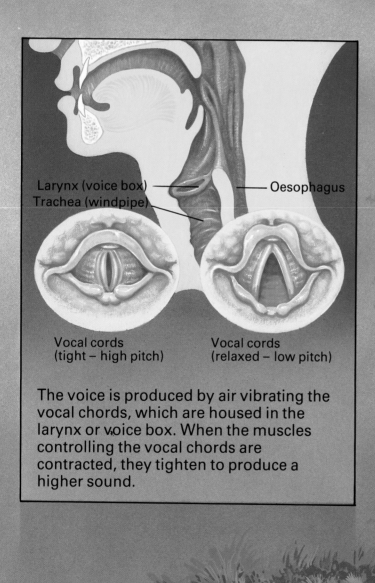

Larynx (voice box) — Oesophagus
Trachea (windpipe)

Vocal cords (tight – high pitch)   Vocal cords (relaxed – low pitch)

The voice is produced by air vibrating the vocal chords, which are housed in the larynx or voice box. When the muscles controlling the vocal chords are contracted, they tighten to produce a higher sound.

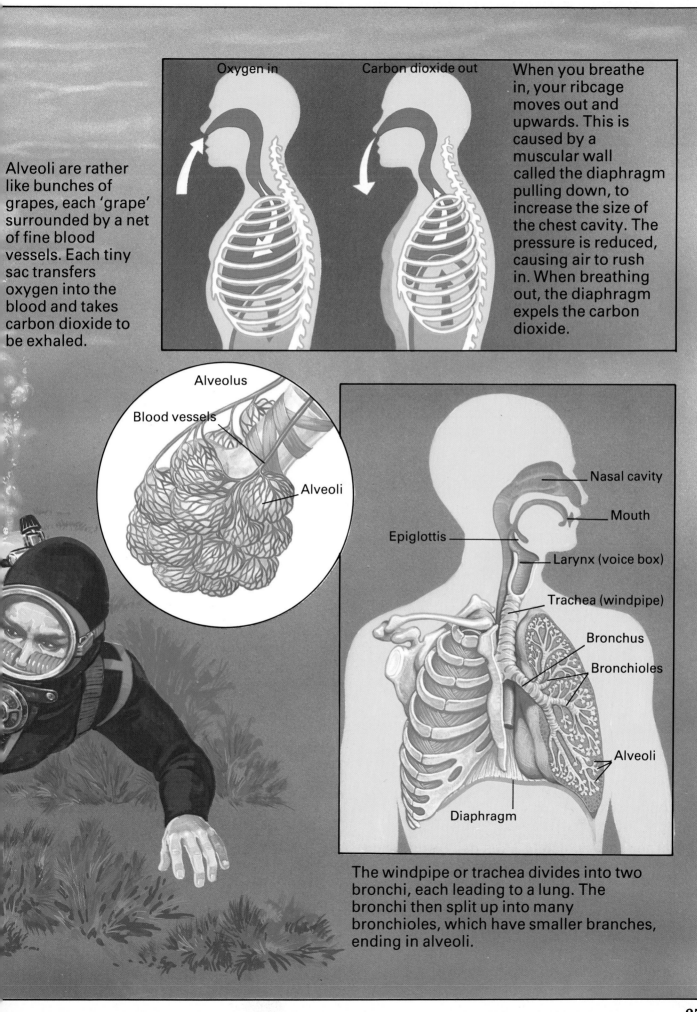

Alveoli are rather like bunches of grapes, each 'grape' surrounded by a net of fine blood vessels. Each tiny sac transfers oxygen into the blood and takes carbon dioxide to be exhaled.

Oxygen in

Carbon dioxide out

When you breathe in, your ribcage moves out and upwards. This is caused by a muscular wall called the diaphragm pulling down, to increase the size of the chest cavity. The pressure is reduced, causing air to rush in. When breathing out, the diaphragm expels the carbon dioxide.

Alveolus

Blood vessels

Alveoli

Nasal cavity

Mouth

Epiglottis

Larynx (voice box)

Trachea (windpipe)

Bronchus

Bronchioles

Alveoli

Diaphragm

The windpipe or trachea divides into two bronchi, each leading to a lung. The bronchi then split up into many bronchioles, which have smaller branches, ending in alveoli.

What fuels your body?

The digestive system is like a chemical factory, it processes the food we eat into simple compounds that can be absorbed into the blood stream. This process starts in the mouth where the teeth break the food into small particles and mix it with saliva.

The food is then squeezed, by muscular action, down the oesophagus into the stomach where it is mixed with digestive juices into a liquid. A valve in the stomach releases this mixture into the small intestine, which is lined with millions of small hair-like projections called villi. By the time this watery mixture, called chyme, reaches the large intestine, it has travelled 7 metres from the stomach.

A balanced diet contains three basic kinds of food – protein, fats and carbohydrates. These foods provide us with energy which is measured in calories. The number of calories each person needs a day depends on how active they are, but it is normally between 2–3,000.

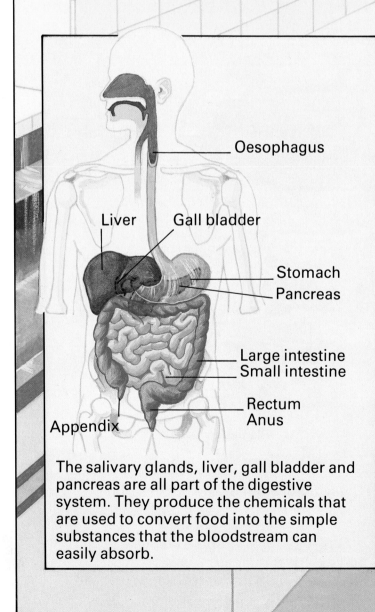

Oesophagus

Liver   Gall bladder

Stomach
Pancreas

Large intestine
Small intestine

Rectum
Anus

Appendix

Villi

Artery   Vein

Lymph vessels

Muscle

The salivary glands, liver, gall bladder and pancreas are all part of the digestive system. They produce the chemicals that are used to convert food into the simple substances that the bloodstream can easily absorb.

Tongue

Teeth

Salivary glands

Oesophagus

The teeth break the food into small pieces which are mixed with saliva from the salivary glands.

Vitamin A is needed for a healthy skin and respiratory system, and also for the eyes.

Vitamin B helps the body to release energy and in the production of red blood cells.

Vitamin C makes a substance which binds body tissue together, and helps us resist germs.

Vitamin D builds strong bones and teeth. It is produced by skin exposed to sunlight and found in some foods.

Vitamin E is important for reproduction. It is present in most green vegetables.

Vitamin K is used to produce a substance which is used for clotting the blood. It is made by bacteria in our intestines and is also present in green vegetables.

In order to remain healthy, the body needs a regular supply of certain vitamins. These vitamins provide the material from which the body produces new cells.

What are chemical messengers?

Robert Wadlow, born in America in 1918, grew to be 2.72 m tall, due to the over activity of a tiny gland in the base of his brain, called the pituitary gland. This gland controls the production of a hormone that is responsible for the body's growth.

The pituitary gland is part of the endocrine system, which produces hormones. These hormones are, in fact, chemical messengers which control many of the body processes. The whole system is controlled by a part of the brain called the hypothalamus, which governs the action of the pituitary gland.

In a dangerous situation your body automatically prepares itself for action. Your breathing rate increases, and your heart beats faster to supply extra blood. The liver increases sugar production to provide energy for the muscles which propel you away from the danger.

Here a giant and a dwarf are compared with a man of average height (5'9" – 1.77 m). Both of these conditions are caused by a malfunction of the gland controlling the growth hormones. In some cases these conditions can now be successfully treated.

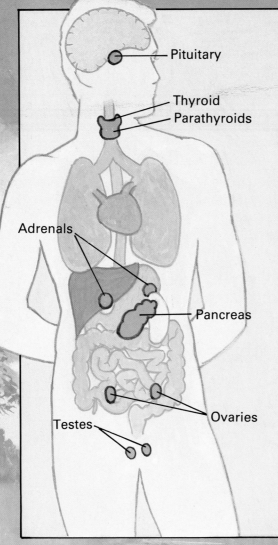

Pituitary
Thyroid
Parathyroids
Adrenals
Pancreas
Ovaries
Testes

The Endocrine System

The pituitary gland is the master gland which controls other glands and makes many different hormones. It produces hormones that affect growth, egg production, and the birth process.

The thyroid produces thyroxine, a hormone which regulates the metabolism or speed of the body's processes – for example, the breathing rate.

The pancreas makes insulin, which controls the level of sugar in the blood.

The adrenal glands control fluid levels and minerals in the blood, and regulate the heartbeat.

The ovaries and testes produce sex hormones. These control our sexual characteristics.

Are you electric?

The nervous system of the body is like the communications system of a large country. It not only transmits information about what is happening within the country, but also concerning events in the world around.

The sense organs – sight, hearing, touch and smell – are like radar, satellites or foreign correspondents; they gather information about the outside world so that the body can decide how to act. The internal nervous system transmits this information to every cell, regulating the activities of muscles, glands and organs.

Pain is like an alarm system, it warns you when your body is being damaged. Some people choose to ignore these warnings. You may hear athletes discussing the pain barrier, which is a warning from the muscles that they are working too hard. The Indian fakir below is ignoring the warnings.

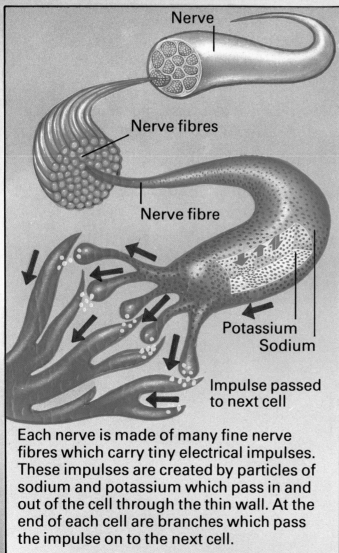

Nerve

Nerve fibres

Nerve fibre

Potassium
Sodium

Impulse passed
to next cell

Each nerve is made of many fine nerve fibres which carry tiny electrical impulses. These impulses are created by particles of sodium and potassium which pass in and out of the cell through the thin wall. At the end of each cell are branches which pass the impulse on to the next cell.

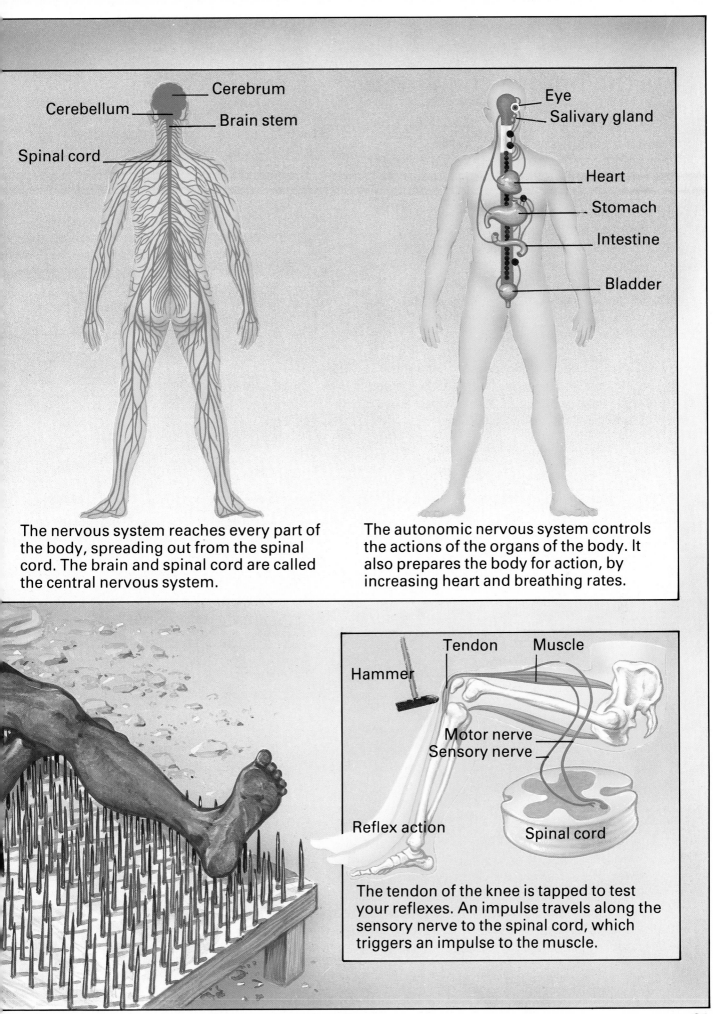

Cerebrum

Cerebellum

Brain stem

Spinal cord

The nervous system reaches every part of the body, spreading out from the spinal cord. The brain and spinal cord are called the central nervous system.

Eye

Salivary gland

Heart

Stomach

Intestine

Bladder

The autonomic nervous system controls the actions of the organs of the body. It also prepares the body for action, by increasing heart and breathing rates.

Hammer

Tendon

Muscle

Motor nerve

Sensory nerve

Reflex action

Spinal cord

The tendon of the knee is tapped to test your reflexes. An impulse travels along the sensory nerve to the spinal cord, which triggers an impulse to the muscle.

Is the brain a computer?

The human brain is a mass of greyish-pink jelly, weighing only about 1,400 g. Yet this grapefruit-sized organ is more complex and versatile than the most powerful computer.

The main part of the brain, the cerebrum, is divided into two hemispheres. These are covered with folds and creases which give a greater surface area without increasing the size of the brain. Another part of the brain is called the cerebellum. This controls the body's movements and co-ordination. In primitive animals, like fish or reptiles, this part is larger than the cerebrum.

It is the cerebrum that is responsible for intelligence and thought as we understand it. Although we know very little about the actual process of thought, we do know which areas of the brain perform certain functions.

The human brain cannot match the computer's speed at pure calculation, as a computer can perform thousands of calculations per second. The human brain, however, is much more versatile than any computer. It can run the complex body machine automatically, whilst engaged in activities as varied as playing tennis or even designing a computer.

Blue whale

Stegosaurus

Elephant

Chimpanzee

Dolphin

Man

Intelligence cannot be measured by the size of the brain. If this was the case, the sperm whale would be the most intelligent animal with a recorded brain weight of 9.2 kg. However, it seems that the proportion of brain to body weight may be a guide to intelligence. After Man, the next most intelligent animals, if measured this way, are dolphins and chimpanzees.

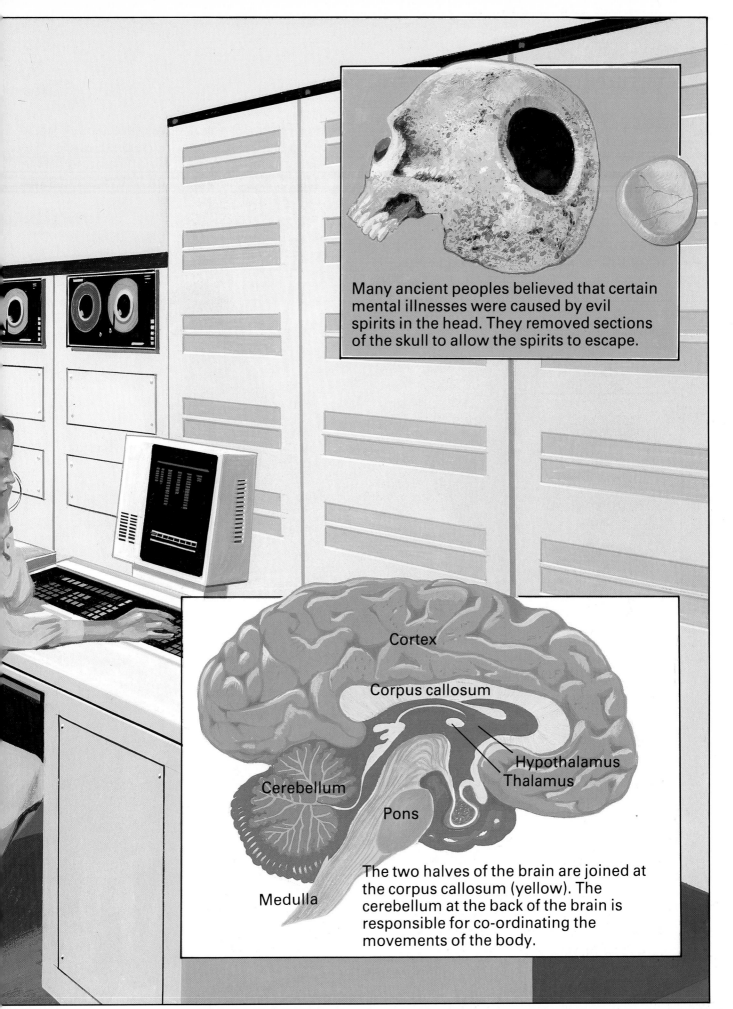

Many ancient peoples believed that certain mental illnesses were caused by evil spirits in the head. They removed sections of the skull to allow the spirits to escape.

Cortex

Corpus callosum

Hypothalamus
Thalamus

Cerebellum

Pons

Medulla

The two halves of the brain are joined at the corpus callosum (yellow). The cerebellum at the back of the brain is responsible for co-ordinating the movements of the body.

What has rods and cones?

The human eye is like a camera. Its pupil is the lens which focuses the image. The iris around the lens controls the amount of light let in, like the aperture of a camera. But instead of film, the eye has a retina made up of millions of light-sensitive cells connected to the brain by the optic nerve. These light-sensitive cells are called rods and cones, and a healthy retina has about 130 million of them. The cones are responsible for colour vision and work only in fairly bright light. When there is little light, only the rods are used, which is why things appear to be grey in dim lighting.

In the modern world, we rarely need good long-distance sight. In the past, and amongst tribes that rely on hunting for game, like the Bushman below, long sight is essential for survival.

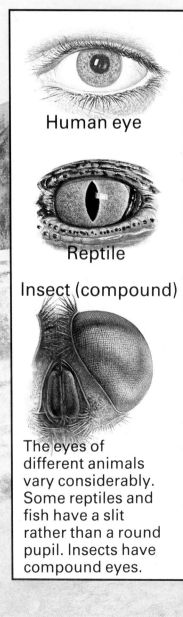

Human eye

Reptile

Insect (compound)

The eyes of different animals vary considerably. Some reptiles and fish have a slit rather than a round pupil. Insects have compound eyes.

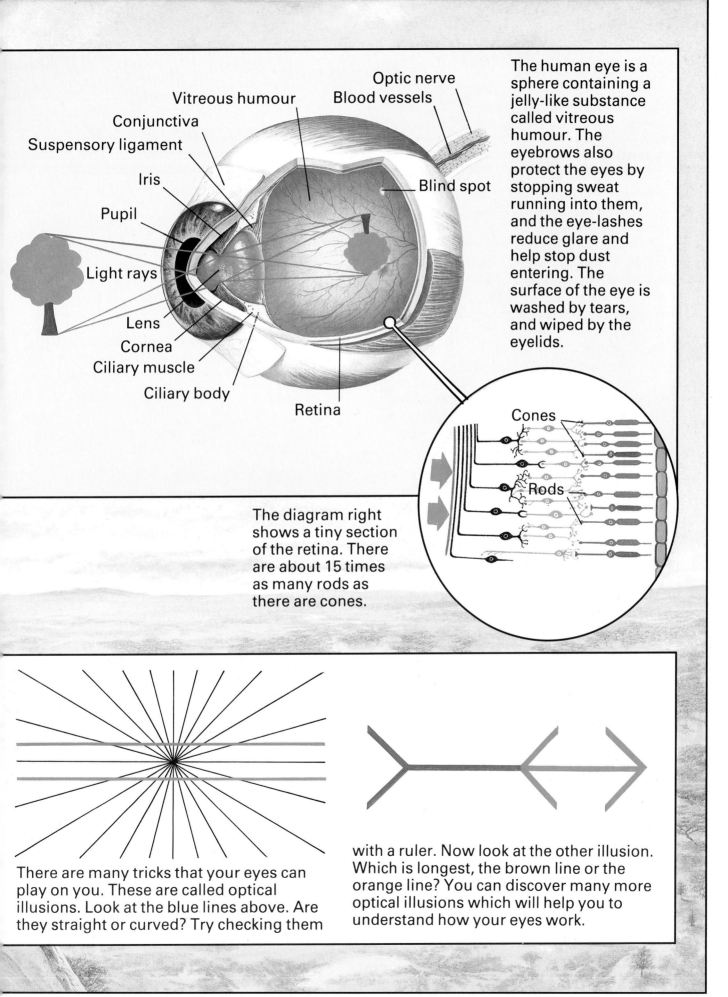

Optic nerve
Blood vessels

Vitreous humour

Conjunctiva

Suspensory ligament

Iris

Pupil

Blind spot

Light rays

Lens

Cornea

Ciliary muscle

Ciliary body

Retina

The human eye is a sphere containing a jelly-like substance called vitreous humour. The eyebrows also protect the eyes by stopping sweat running into them, and the eye-lashes reduce glare and help stop dust entering. The surface of the eye is washed by tears, and wiped by the eyelids.

Cones

Rods

The diagram right shows a tiny section of the retina. There are about 15 times as many rods as there are cones.

There are many tricks that your eyes can play on you. These are called optical illusions. Look at the blue lines above. Are they straight or curved? Try checking them

with a ruler. Now look at the other illusion. Which is longest, the brown line or the orange line? You can discover many more optical illusions which will help you to understand how your eyes work.

Do ears help you balance?

We are almost constantly bombarded with sound, and yet we have an ability to focus our listening. It is possible to be almost unaware of the roar of passing traffic, and yet hear the tiny sounds of a key in the door. This selective hearing is the brain editing the background noise, while remaining constantly alert for any important sounds.

Sounds occur when a vibrating object sets the molecules of the air in motion. They move in waves – the closer the sound waves, the higher the sound. Although our hearing range is rather limited compared with many other animals, it is very sensitive. This is due to the brain, which has the job of comparing the sounds received with those in the memory.

When the sound waves reach the outer ear, they are funnelled into the eardrum, which vibrates. The vibrations are magnified by three tiny bones in the middle ear. They are sent through the fine membrane of the oval window to the inner ear, where they pass along the cochlea to the organ of Corti, which transmits sound as electrical impulses through the auditory nerve to the brain.

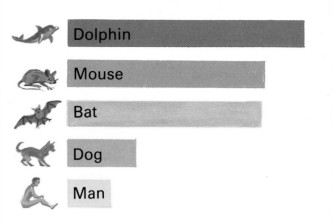

Dolphin

Mouse

Bat

Dog

Man

Although our hearing is well adapted to the range of human speech, it is limited in range when compared with other animals.

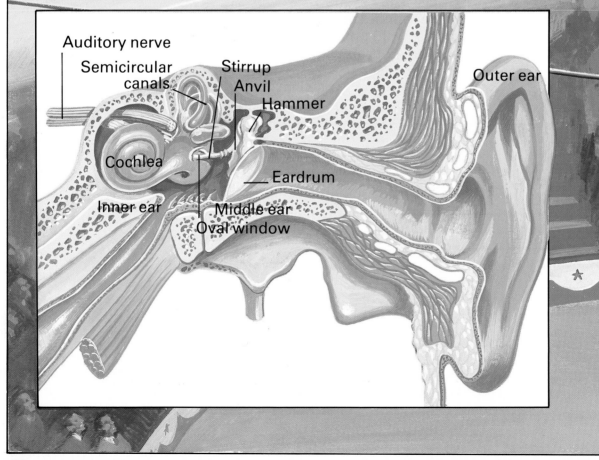

Auditory nerve
Semicircular canals
Stirrup
Anvil
Hammer
Outer ear
Cochlea
Eardrum
Inner ear
Middle ear
Oval window

Our sense of balance is a very complex thing; it allows us to 'feel' the stability of our body's position, even if our eyes are closed. The mechanism of balance is in the inner ear. The three semicircular canals inform the brain if the head is tilted, and the brain then makes continuous adjustments to the muscles in order to compensate for the pull of gravity.

Ear trumpet

Earpiece

Battery powered hearing aid

Microphone
Battery case

Volume control

Modern hearing aid

The first hearing aids were funnels which collected sound and directed it into the ear. The next development was a microphone to pick up the sound, and convert it into an electrical signal which powered a tiny loudspeaker in the ear.

What has 10,000 buds?

Imagine how boring our food would be if it all tasted the same. Without variety of taste, food would not be a pleasure, but purely nourishment for the body. Yet the pleasure we get from food is not just taste; if you hold your nose it is not easy to tell whether you are eating apple or potato. This is because the flavour of food is a combination of both taste and smell.

Our sense of smell is far less developed than in many animals. We do not use it for finding food or avoiding enemies. This is probably due to our upright stance as, unlike a dog or any other four-footed creature, our nose is very far from the ground. Instead of smell, man has relied on his sight and hearing for hunting and to warn him of danger. Smell and taste are detected by special cells on the tongue and in the nasal cavity. These cells have tiny hairs which project through pores in the surface. The hairs react to minute particles in the air or in the saliva, and send a message through the nervous system to the brain.

The wine-taster below has very well developed senses of taste and smell. He is able to recognise thousands of different wines, by remembering their characteristic 'flavour'.

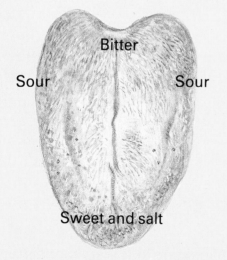

Bitter

Sour

Sour

Sweet and salt

The tongue is covered with cells called 'taste buds'. These cells are located in tiny trenches on the surface. The taste buds cover the tongue, but some parts are more sensitive to specific tastes than others.

Sensory hair

Taste cell

Nerve fibre

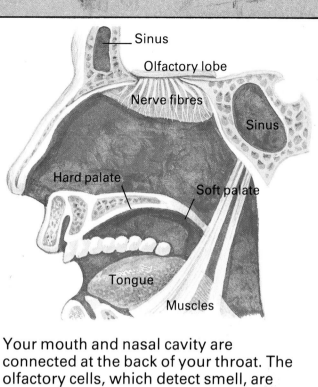

Sinus

Olfactory lobe

Nerve fibres

Sinus

Hard palate

Soft palate

Tongue

Muscles

Your mouth and nasal cavity are connected at the back of your throat. The olfactory cells, which detect smell, are located in the top of the nasal cavity.

This test will show you how important smell is, when tasting food. Try to find things that have a similar texture, like apples and potato, or bread and cake. Close your eyes and hold your nose; you will be surprised at how difficult it is to tell the difference.

Why do we chew?

The mouth is the first stage in the process of digestion. This is where the food we eat is broken down into smaller pieces and mixed with saliva for swallowing.

The tools we use for this process are our teeth. Human teeth are a versatile tool kit comprising incisors for cutting, canines for tearing and molars for grinding. Our teeth are the product of evolution and are a good indication of the mixed diet of our ancestors. Many animals have a very specialised diet, some, like the anteater, have no need for teeth at all, whilst others have teeth to suit their specialised eating habits.

If you care for your teeth, they will last throughout your life. Perhaps the man below, holding back a powerful truck with his teeth, is not setting a good example. He is John Massis, the man with the strongest teeth in the world. He has towed trucks, trains and even held down a helicopter with his teeth. If you want your teeth to last, this is not to be recommended.

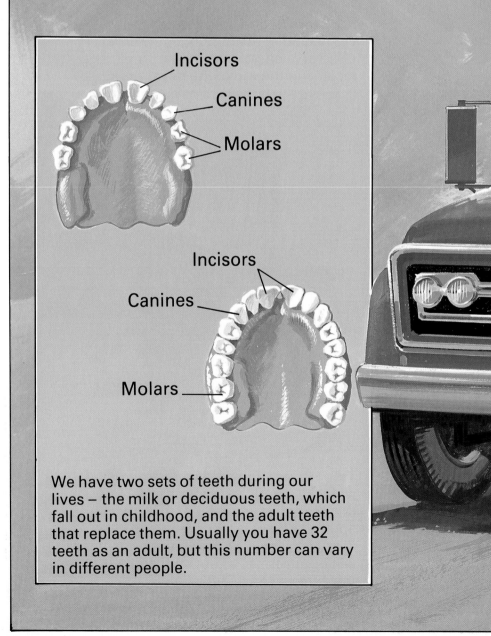

Incisors

Canines

Molars

Incisors

Canines

Molars

We have two sets of teeth during our lives – the milk or deciduous teeth, which fall out in childhood, and the adult teeth that replace them. Usually you have 32 teeth as an adult, but this number can vary in different people.

Animals have teeth to suit their eating habits and behaviour. The beaver has strong, sharp incisors to gnaw wood for food and building. Lions use their dagger-like teeth for tearing meat. The herbivores have large, flat molars to grind the tough vegetation of their diet.

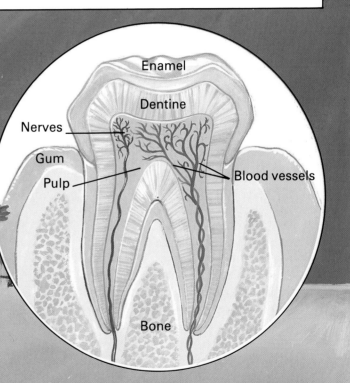

Enamel

Dentine

Nerves

Gum

Pulp

Blood vessels

Bone

A tooth is composed of a layer of hard enamel covering a softer bone-like substance called dentine. Inside the tooth is soft pulp containing the nerves and blood vessels. The tooth roots are held in the jawbone by a strong glue-like material.

What gives 2 sq m of protection?

We may not have the armoured protection of a crocodile, or the dramatic colouring of the tiger, but our skin gives us both protection and information.

The skin has been called the largest organ of the body. In a normal adult it covers an area of 1.5–2 sq m and performs a wide variety of jobs. It protects our inner organs from the environment, providing a tough barrier against bacteria and physical damage. It also maintains our temperature, increasing blood flow and producing sweat when we are hot.

The skin is also a communications network, containing cells sensitive to heat, cold, pressure, touch and pain.

Many primitive peoples have used their skin for decoration, by painting or marking it in other ways. The people below are the Nuba of Central Africa, who decorate their skins by painting or even more permanently by scarring it with patterns of small cuts. Although this must be very painful, it is done purely for the sake of art and beauty.

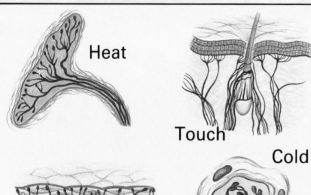

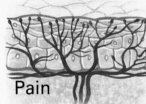

Heat

Touch

Cold

Pain

The inner layer of skin, or dermis, contains the cells responsible for our sense of touch. These cells separately register sensations of heat, cold, pain, touch and pressure. Our reaction to touching an object is the result of a combination of these sensations.

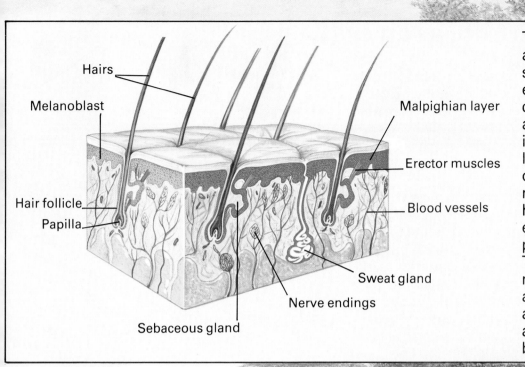

Hairs

Melanoblast

Hair follicle

Papilla

Sebaceous gland

Nerve endings

Sweat gland

Malpighian layer

Erector muscles

Blood vessels

The diagram shows a small section of skin. Almost the entire body is covered in hair, although much of it is very fine and fair. In a situation of cold or perhaps fear, tiny muscles cause each hair to become erect and 'goose pimples' to appear. This is common to mammals as it traps a layer of warm air, and makes the animal appear bigger.

Like the tread on a car tyre, the pattern of ridges on the skin of our hands improves grip. As every fingerprint is unique, it is a useful means of identification.

Did you start life as an egg?

The human body is an organism composed of billions of individual cells. At the moment when life begins, the egg is fertilised by a male sperm. This future human being is then a single cell. Yet this single cell contains all the information necessary to build and organise a complete person.

The moment when the egg and sperm combine is called conception. After about 30 hours, the egg divides into two cells and then four, eight, sixteen, each time doubling the number of cells. Soon there is a cluster of cells like a tiny bunch of grapes. This settles into the womb and continues to multiply and grow.

As it grows larger, the embryo is nourished through a tube called the umbilical cord, and by about six weeks, begins to develop tiny arms and legs, although it is still only a few centimetres long. From this time the developing baby is called a foetus. After six months it is completely formed, and at nine months it is ready to be born.

At the moment of birth you have already been growing for about nine months, but you still have many years of growth and development before reaching maturity.

1–9 months

2–4 years

20–30 years

40–50 years
Middle age

60–75 years
Grandparent

During its development, the foetus is protected in the watery liquid of the amniotic sac, and fed through the umbilical cord. There are now ways of seeing the developing foetus inside the womb. A picture can be created by using sound waves, which are harmless, but can show enough detail even to see the tiny beating heart.

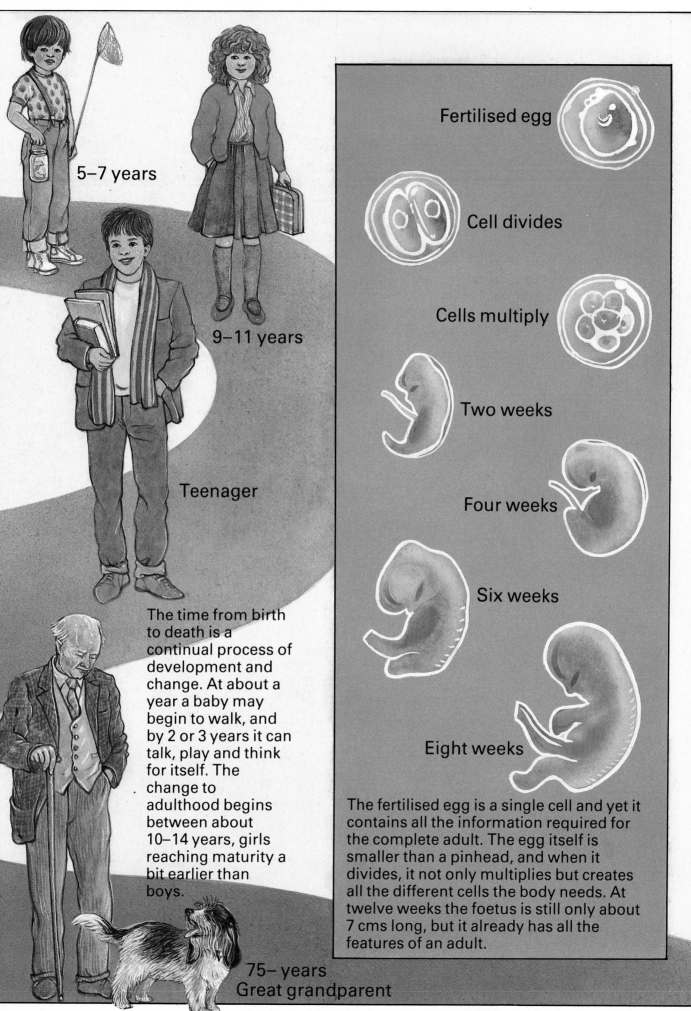

5–7 years

9–11 years

Teenager

The time from birth to death is a continual process of development and change. At about a year a baby may begin to walk, and by 2 or 3 years it can talk, play and think for itself. The change to adulthood begins between about 10–14 years, girls reaching maturity a bit earlier than boys.

75– years
Great grandparent

Fertilised egg

Cell divides

Cells multiply

Two weeks

Four weeks

Six weeks

Eight weeks

The fertilised egg is a single cell and yet it contains all the information required for the complete adult. The egg itself is smaller than a pinhead, and when it divides, it not only multiplies but creates all the different cells the body needs. At twelve weeks the foetus is still only about 7 cms long, but it already has all the features of an adult.

45

Do you look like your parents?

All the instructions for building the complete human being are carried in the single egg from which life begins. These instructions not only carry the basic information to produce and organise cells for bone, nerves, muscles and brain, but also the specific characteristics of both parents. These may be colour of eyes or hair, but can also be special abilities, like intelligence or artistic skills.

Every cell of your body contains a 'copy' of the original instructions and, in addition to giving you your own personal characteristics, these will be passed on to your children.

This family o
grandparents
grandchildre
hair and carri
her tee-shirt.
also has red h
Other membe
'red' gene an
hair, the 'red'
their children

Grandparent

Grandparent

These two girls look very alike – they are identical twins. This happens when two embryos are formed from a single egg. As the characteristics of each person are controlled by the instruction code of the genes, these girls have exactly the same characteristics because they have the same genes.

Parent

Parent

Child

Child

ch is composed of
ildren, and
ndmother has red
ed' genes, shown on
r two daughters
o does a grandson.
family have one
h they have dark
be passed on to

Gregor Mendel was an Augustinian monk who lived in Austria in the 19th century. In 1865, he explained the laws of inheritance that he had learned from growing pea plants in the garden of his monastery. It was not until 1953, almost a century later, that we began to understand how inherited characteristics are controlled by the genes.

Parent

Parent

Child

Child

Child

Why are people different colours?

The origin of the skin-colour of each race is due to the climate in which its ancestors evolved. The production of melanin protects the delicate underlying tissues from the damaging effects of strong sunlight. This works in a similar way to the use of tinted sunglasses for protecting the eyes from the sun's glare. In this way people with origins in very hot, sunny climates have dark skins and those from the cooler temperate regions are paler.

Evolution is a continuous process. Perhaps the millions of black people who have now lived in cooler areas for many generations will gradually become paler. The opposite could also happen in parts of Australia, where fair-skinned people have moved to a hotter climate.

In the last few hundred years, people have moved in vast numbers around the earth. They have settled in new areas and married people of different origins. As this process continues, the differences between races will gradually begin to disappear.

Hair varies in colour and texture according to racial origins. Whether hair is straight, wavy or curly depends on the shape of each individual hair. Straight hair is round in section, wavy hair is oval and curly hair is flat.

Caucasoid (white)

Mongoloid (yellow)

Negroid (black)

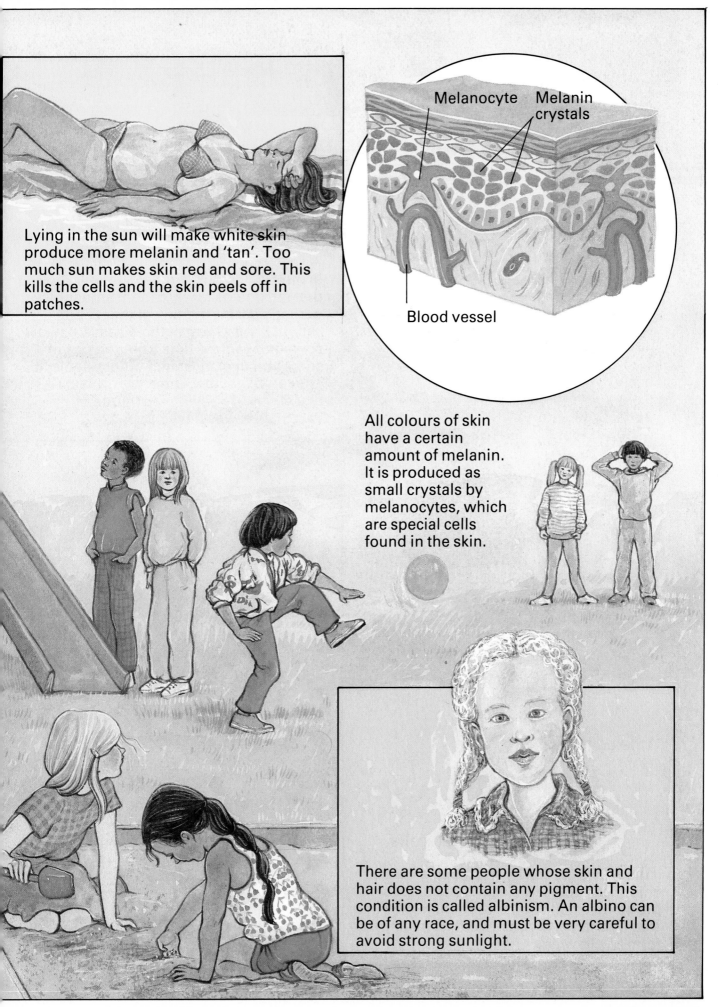

Lying in the sun will make white skin produce more melanin and 'tan'. Too much sun makes skin red and sore. This kills the cells and the skin peels off in patches.

Melanocyte Melanin crystals

Blood vessel

All colours of skin have a certain amount of melanin. It is produced as small crystals by melanocytes, which are special cells found in the skin.

There are some people whose skin and hair does not contain any pigment. This condition is called albinism. An albino can be of any race, and must be very careful to avoid strong sunlight.

How does the body defend itself?

In the event of danger or invasion, it is the body's defence system that is alerted. This defence system is an army of white blood cells, that quickly arrive on the scene to deal with invading bacteria.

When a white cell encounters a germ it engulfs it in its liquid body. Chemicals in the white cell then destroy the germ and the remains are expelled.

Some types of germ cannot be attacked by these white cells. This is because they have a protective coating around them. In this case, a chemical called an antibody will first destroy the coating before the white cell attacks. Each type of germ may need a different antibody. These are either produced by the body or absorbed from the mother before birth. They can also be developed by artificially introducing a small amount of a mild strain of the bacteria.

In the 18th century Edward Jenner noticed that milkmaids rarely caught smallpox, perhaps because many of them had caught a mild form of the disease, called cowpox. He drew fluid from the cowpox sores and injected it into other people. It appeared that they too were protected, and so the life-saving technique of vaccination was developed.

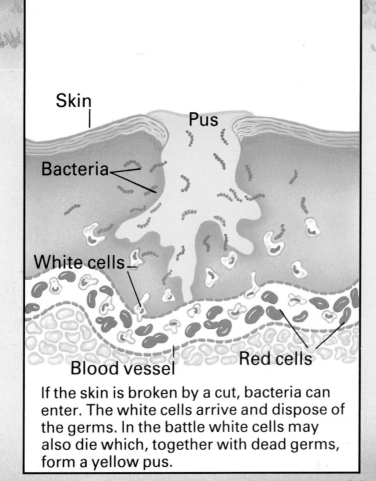

Skin

Pus

Bacteria

White cells

Blood vessel

Red cells

If the skin is broken by a cut, bacteria can enter. The white cells arrive and dispose of the germs. In the battle white cells may also die which, together with dead germs, form a yellow pus.

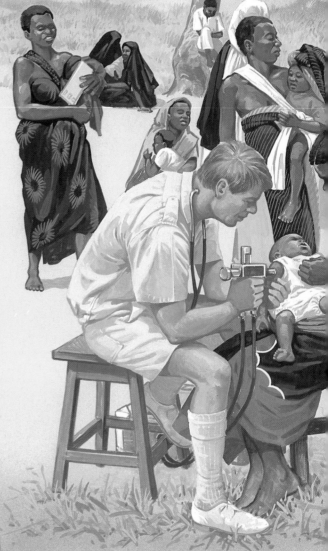

In many parts of the world, certain diseases were once very common. In the developing countries, huge numbers of people died from diseases which are now rare in the west. In recent years huge programmes of mass vaccination have now been carried out, gaining control of many of these diseases.

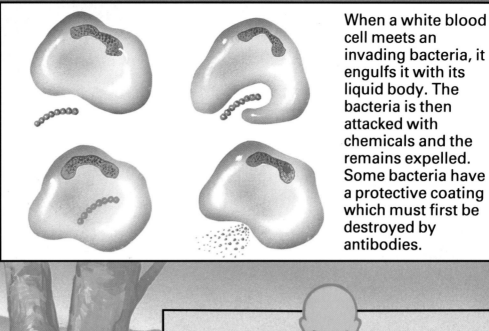

When a white blood cell meets an invading bacteria, it engulfs it with its liquid body. The bacteria is then attacked with chemicals and the remains expelled. Some bacteria have a protective coating which must first be destroyed by antibodies.

Lymph vessels   Lymph node

The lymphatic system is a network of vessels, rather like the blood system. One of its roles is the supply and removal of the fluids in the body. It also acts as a transport system for white blood cells.

Can people be repaired?

Although the body has a marvellous ability to fight off invading germs and to repair itself after injury, there are times when medical science must come to the rescue.

In recent years we have seen major advances in every aspect of medicine, from the treatment of common diseases to the transplanting of hearts, lungs and kidneys.

One of the most amazing recent developments is the replacement of parts of the body with synthetic substitutes. This was originally limited to the skeletal framework, with operations like hip replacements becoming almost routine. However, we are now approaching an era with even more far-reaching possibilities. In 1982, in the United States, the first artificial heart was implanted into a human body.

The modern operating theatre is kept sterile (germ free) and at a constant temperature. Major operations require a large team of specialist doctors and nurses, who perform the operation and monitor the patient's condition, using a wide range of electronic equipment.

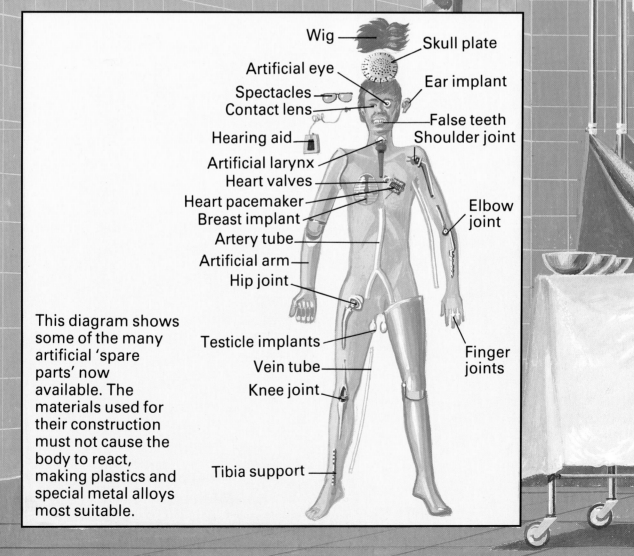

Wig — Skull plate
Artificial eye — Ear implant
Spectacles
Contact lens
False teeth
Shoulder joint
Hearing aid
Artificial larynx
Heart valves — Elbow joint
Heart pacemaker
Breast implant
Artery tube
Artificial arm
Hip joint
Testicle implants — Finger joints
Vein tube
Knee joint
Tibia support

This diagram shows some of the many artificial 'spare parts' now available. The materials used for their construction must not cause the body to react, making plastics and special metal alloys most suitable.

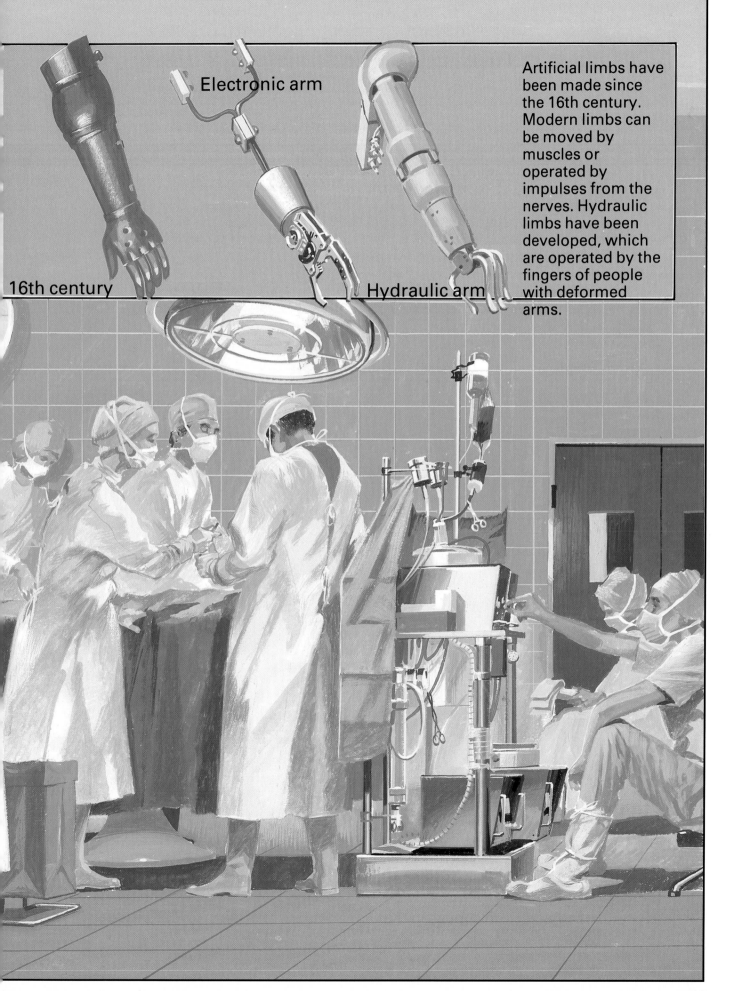

Electronic arm

16th century

Hydraulic arm

Artificial limbs have been made since the 16th century. Modern limbs can be moved by muscles or operated by impulses from the nerves. Hydraulic limbs have been developed, which are operated by the fingers of people with deformed arms.

Can Man compete with animals?

Man is the dominant creature on the earth. This means that we have control over all other living things. Yet, compared with other animals, we may not seem to have very great physical ability. We cannot run as fast as the cheetah or antelope. We cannot jump as well as a tiger or kangaroo. We cannot swim like a fish or seal. We are not as strong as a gorilla or a lion, and we cannot fly like birds or insects.

Man's success is due to two things – a large and powerful brain, and hands suitable for gripping tools. Using these two things we can build machines that are faster and stronger than any animal.

One of Man's great advantages is the ability to adapt to any climate and environment. People can survive in the hottest deserts and the coldest places on Earth. Yet one of the most difficult places to adapt to is the crowded city. Early Man

We may be very impressed by the ability or our athletes as they compete against each other in sporting events. Yet, compared with other animals, our physical abilities are very limited. Animals can run faster and further, jump higher and longer, and swim faster and deeper.

as a hunter; as time passed people ettled in groups to farm the land. Only fairly recently have we gathered in vast cities of many millions of people.

Do we have supernatural powers?

Man's history is a mixture of science and magic. We have always struggled to understand the world around us and, although our knowledge of science has grown, we may have lost other natural abilities.

Most primitive peoples have a deep belief in man's supernatural powers. These beliefs have generally disappeared in the western world, as science has developed. It is only in recent years that scientists have begun to take an active interest and try to establish if these powers really exist.

There have been many claims for different kinds of supernatural ability. These cover contact with the spirit world, healing by touch, communicating by thought and affecting objects by thought. Some people even claim to be able to 'see' into the future, but there has been very little scientific proof.

Uri Geller was very famous for his unusual powers. He appeared to be able to bend metal just by stroking it gently. Some people claim that metal objects in their homes bent when he appeared on the television.

There are now machines that have been designed to test for ESP (Extra Sensory Perception). The person being tested has to choose symbols matching those selected on a machine they cannot see.

A gathering of people trying to contact the spirits of the dead is called a séance. These were very popular in Victorian times. The séance is led by a medium, who has special powers, and is used by the spirits for communication. Although many people believe in spiritualism, there is little scientific evidence to prove it.

Dowsers claim the ability to locate water or certain minerals by the movement of a forked twig or wire held in their hands.

What can the body achieve?

Many people have special natural abilities, like unusual strength or speed. With training, these abilities can be increased to an almost superhuman level.

The body is a flexible machine, it can be trained for a wide range of very different purposes. Some people develop special skills for sport and others for entertainment. By regular practice your body can become more adapted to suit you, whether you want to be an athlete or a musician.

WARNING

Many things you see, like sword-swallowing or attempting to break tiles with the hand, are extremely dangerous. They can only be achieved after many years of training with expert teachers. You should never attempt them yourself, as they could result in serious injury.

This girl is able to arrange her limbs in positions that most people would find impossible. It may be partly due to a natural ability, but it can also be achieved by practice.

The karate expert below has smashed a stack of roof tiles, using only the edge of his hand. Karate is an ancient Japanese martial or fighting art, needing many years of dedicated training.

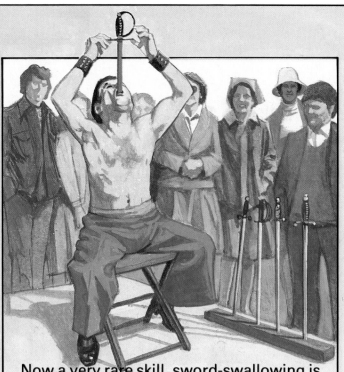

Now a very rare skill, sword-swallowing is a dangerous occupation. If you know a little about human anatomy you will understand why.

Weightlifting is an Olympic sport, but it is also a way of increasing muscle size. Some people enter competitions where the body's shape, rather than strength, is judged.